Securing America's Future

The Vital Role of STEMM Education in National Defense and Innovation

Dr. Judy Staveley

Securing America's Future

The Vital Role of STEMM Education in National Defense and Innovation

©2024

All rights reserved. No part of this publication may be reproduced or transmitted in any form or by any means, electronic or mechanical, including photocopying, recording, or by any information storage and retrieval system, without the prior written permission from the authors, except by reviewers who may quote brief excerpts in connection with a review in scholarly publications, newspaper, magazine or electronic publication. Contact the author information on foreign rights.

1st edition

ISBN: 9798344542546

Printed and Written in the United States

Acknowledgements

To my family who have served in the military and fought for our freedom of speech, education, and national security.

To my husband who has been my biggest support of my academic, scholarly and writing career.

To my kids who love STEMM and support my efforts in promoting STEMM education.

Table of Contents

Preface 5
About the Author Dr. Judy Staveley 7

Chapter 1: STEMM – The Foundation of Progress 8

Chapter 2: STEMM Education for National Security 12

Chapter 3: The Founding Fathers and the Vision for Education 17

Chapter 4: Beyond the Classroom: STEMM in the Real-World 23

Chapter 5: Partnerships in Education and Innovation 30

Chapter 6: Workforce Development for a Modern Society 37

Chapter 7: The Role of Scientific Societies 44

Chapter 8: Biotechnology The Frontier of Emerging Technologies 55

Chapter 9: Emerging Trends and the Future of STEMM 64

Chapter 10: Securing the Future through STEMM Education and Innovation 73

Glossary 81
References 83

Preface

STEMM education which encompasses Science, Technology, Engineering, Mathematics, and Medicine has become central to national security in the 21st century.

This book aims to bridge the gap between historical educational values, such as those cherished by our Founding Fathers and the contemporary need for innovation and defense through an educational lens.

From the emergence of biotechnology to the rise of partnerships and beyond the classroom this journey explores how we can cultivate a future workforce capable of safeguarding our country.

STEMM

S - Science

T - Technology

E - Engineering

M - Mathematics

M - Medicine

About the Author
Dr. Judy Staveley

Dr. Judy Staveley is a recognized scholar who is an advocate for STEMM education. With a distinguished career as a Senior Scientist and Subject Matter Expert she has a deep commitment to promoting education, workforce development, and scientific innovative practices. Dr. Staveley has devoted her career to empowering students and professionals alike in the pursuit of national security through science and technology.

Through her various roles in academia, government, and the private sector, Dr. Staveley has established herself as a leading voice in shaping the future of STEMM.

For more on Dr. Staveley's journey and work, visit her website at www.DrJudyStaveley.com

Chapter 1
The Foundation
and
Progress of STEMM

Science, Technology, Engineering, Mathematics, and Medicine (STEMM) are the cornerstones of modern society, driving innovation, growth, and national security. Without the innovative advancements brought through STEMM initiatives and competitions, the world as we know it would be vastly different. Each of these disciplines touch every aspect of our daily lives. Cell phones that we use to communicate, to the medical treatments that save lives, medical diagnostics, vaccines, and computer technology. This chapter will introduce you to STEMM in a simple and approachable way, so that you as a reader can enjoy learning about each of these fields and how they work together to support the society's progress.

STEMM is the engine that powers nearly every facet of our world. Science forms the foundation of knowledge, research and development, while Technology applies that knowledge to create tools and systems. Engineering brings those tools to life by designing and building the infrastructure and devices we rely on daily. Mathematics underpins all these areas by providing the language of problem-solving, enabling precision and logic in everything from software development to space exploration. Lastly, Medicine ensures that scientific breakthroughs are translated into health advancements that improve and save lives.

One of the most visible examples of STEMM's impact on society occurred during the 2020 COVID-19 pandemic. The collaboration between scientists, engineers, and medical professionals led to the rapid development of vaccines, which played a critical role in controlling the virus and protecting lives.

The process of vaccine development showcased the intersection of science, technology, and medicine. Research into the virus's structure (science), the creation of mRNA technology (technology), and the application of that research in clinical settings (medicine) all worked together to produce life-saving vaccines (Jackson et al., 2021).

Additionally, the 2020 pandemic highlighted the importance of engineering in public health. Engineers designed medical devices, such as ventilators and personal protective equipment (PPE), to help manage the pandemic crisis. In many cases, engineers adapted existing technologies or developed new ones to meet the sudden surge in demand (Nguyen et al., 2020).

Robotics is another key area of STEMM. Robotics was also employed during the pandemic, for the use in hospitals to minimize human contact and reduce the risk of spreading the virus. These robots handled responsibilities like delivering medication and disinfecting hospital rooms to prevent contamination.

In addition to these advancements, mathematics played an essential role in tracking the spread of COVID-19. Mathematical models helped scientists and policymakers understand how the virus was spreading and were able to predict future outbreaks, and supported professionals to allocate resources effectively. These models informed public health decisions and were crucial for managing the pandemic on both national and global levels (Ferguson et al., 2020).

STEMM not only helped manage the 2020 pandemic but also strengthened its connection to national security. The 2020 pandemic exposed numerous vulnerabilities within our public health systems, underscoring the critical role of STEMM innovation in sustaining national defense.

Having the capability to quickly develop and deploy medical countermeasures, manage supply chains for critical health and medical equipment, and implement technological solutions is vital for protecting the nation in times of crisis. As seen during the 2020 COVID-19 pandemic, a strong foundation in STEMM disciplines ensures that the country can respond effectively to emergencies, whether they are related to public health, cyber threats, or military defense.

As you can see, STEMM is not merely a set of academic concepts; it's the engine driving societal progress. From the vaccines that protect us to the technology that connects us, STEMM disciplines shape our everyday lives. Moving forward, the need for skilled professionals in these fields will only grow as we continue to face new challenges and opportunities. As we dive deeper into this book, you'll discover just how vital STEMM is, not only for individual well-being but also for the nation's security and future prosperity.

Chapter 2
STEMM Education for National Security

In today's complex and rapidly changing world, STEMM education—Science, Technology, Engineering, Mathematics, and Medicine is not only foundational for personal and professional development, but it is also essential for national security and defense for the United States. As we explore chapter two, we will learn how essential a well-educated STEMM workforce strengthens our ability to respond to global challenges and additionally it reinforces our national defense on multiple levels. From cybersecurity to technological innovation, the skills acquired through STEMM disciplines are integral to maintaining a competitive edge and ensuring the safety and the stability of the United States.

The Link Between STEMM and National Security
STEMM subjects play a direct role in enhancing the United States' capabilities in defense and national security. Advances in technology, for instance, lead to new defense tools and systems, making it possible to safeguard against emerging threats both domestically and internationally. Cybersecurity is one area where STEMM expertise is indispensable. Having skilled professionals in computer science and engineering defend critical infrastructure from cyberattacks that could disrupt the economy, compromise personal data, or weaken military defenses.

The Department of Homeland Security (2023) stresses that cybersecurity threats have intensified to a critical national security concern, making it essential to have a skilled workforce in digital defense to address these national security risks.

The national security implications of STEMM fields are clear. Without a trained workforce to maintain and develop these systems, the U.S. would be vulnerable to adversaries who exploit technological weaknesses.

The Need for Skilled Professionals

The increasing complexity of global threats highlights the importance of STEMM trained professionals in sectors like healthcare, defense, technology, and public policy. For example, biotechnology and medical sciences have become essential in addressing biological threats, from pandemics to potential bioterrorism. The COVID-19 2020 pandemic demonstrated the vital importance of rapid innovation when responding to public health crises.
Scientists and researchers worked together to develop vaccines, while engineers designed and manufactured medical devices that ensured the continuity of healthcare during unprecedented times (Jackson et al., 2021).

Similarly, the defense sector requires engineers skilled in aeronautics, robotics, and artificial intelligence (AI) to create and maintain advanced military technologies. AI is an emerging field with significant implications for national security. AI powered technologies are now used in surveillance, risk assessment, and even autonomous defense systems, requiring professionals who understand both the ethical and technical aspects of this powerful technology (Goodman, 2022).

The Role of STEMM Education in Sustaining Global Leadership

For the United States to maintain its leadership on the global stage, STEMM education must be prioritized, both to cultivate a workforce that can innovate and to uphold the nation's competitive standing. Other countries are investing heavily in STEMM to gain technological superiority; they have made it a strategic priority to develop and retain skilled workers in STEMM fields. This competition emphasizes the need for the U.S. to invest in education systems that equip students with the skills necessary to thrive in a global economy centered around technological and scientific progress (National Science Board, 2023).

Building a Pipeline of Future Innovators

To sustain national security, it is essential to develop a robust workforce pipeline of future innovators and problem solvers. K-12 science programs and higher education play a pivotal role in introducing students to STEMM concepts early on. This allows young inspiring scientist to develop a passion for the STEMM fields. Public and private sector partnerships can support internships, research opportunities, and mentorship programs that expose students to real-world applications of STEMM disciplines (Smith, 2023). Government programs like the National Science Foundation's (NSF) STEM initiatives offer grants and fellowships that encourage students from diverse backgrounds to pursue careers in STEMM, thereby strengthening the nation's talent pool and national security.

The Future of STEMM Education for National Security and Defense

Looking ahead, the future of national security depends on a workforce that can adapt to and anticipate technological changes. Education in STEMM fields must go beyond theory, equipping students with practical applied skills to address real-world problems and come up with solutions.

National policies and funding that support STEMM education are investments in the country's safety, resilience, and capacity for innovation. With the increasing complexity of global threats, STEMM's role in defending the United States will only expand, making it crucial for leaders, educators, and policymakers to support and enhance STEMM education at every level.

STEMM education is more than an academic pursuit; it is a strategic asset for national security. In an age of rapid technological advancement, the need for skilled workforce professionals in science, technology, engineering, mathematics, and medicine cannot be overstated. As we look to the future, the United States must continue to invest in and prioritize in STEMM education to protect its position as a global leader, uphold its values, and ensure the safety and security of its people.

Chapter 3
The Founding Fathers and The Vision for Education

The Founding Fathers of the United States recognized that a strong and well-informed society was essential for the nation's prosperity and security. Historical figures like General George Washington, Benjamin Franklin, and later, President Abraham Lincoln, held progressive views on education, science, and innovation, establishing a foundation that continues to shape American values today. Chapter three explores how their vision for education and their fascination with science underscored the importance of knowledge for securing the nation's future.

George Washington
A Leader with an Eye on Science and Education

General and President George Washington, often celebrated for his military and political leadership, was also deeply committed to the advancement of knowledge and education. Washington believed that an educated public was crucial to maintaining liberty and securing the nation's prosperity. He supported the creation of educational institutions that could foster learning and promote scientific advancement. In his Farewell Address, he underscored that knowledge, alongside morality, is a critical pillar for a functional government, expressing a vision where education and civic responsibility are interlinked (Washington, 1796).

Washington's interest extended into agriculture and technology. George Washington believed that these scientific fields were vital to the nation's economy during his time.

George Washington practiced innovative agricultural techniques on his Mount Vernon estate, experimenting with crop rotation and soil conservation, which were early examples of science applied to daily life. His forward-thinking approach illustrates the belief that science could be harnessed to benefit society, a principle that resonates in today's STEMM fields (Smith, 2023).

Benjamin Franklin
The Scientist and Inventor

Benjamin Franklin was perhaps the most scientifically minded of the Founding Fathers, earning recognition as an inventor, philosopher, and diplomat. Franklin's contributions to science, particularly in electricity, transformed understanding in his time and demonstrated his dedication to applying scientific principles for the common good. He conducted groundbreaking experiments, including the famous kite experiment, which provided critical insights into electricity and laid the groundwork for later advancements.

Franklin's passion for science extended into his vision for education. He helped establish the University of Pennsylvania, one of America's earliest universities, with a curriculum that emphasized both the arts and sciences. Franklin believed that education should be practical, equipping students with skills that would directly benefit society. His focus on applied science and education reflected his belief that knowledge could be a transformative force, directly linking his vision to national progress and security (Franklin, 1749).

Thomas Jefferson
Education as a Pillar of Democracy

Thomas Jefferson, the principal author of the Declaration of Independence, was another advocate for the transformative power of education. A polymath, Jefferson was well-versed in fields like architecture, mathematics, and botany. He viewed education as the key to empowering citizens to participate actively in democracy and resist tyranny. Jefferson founded the University of Virginia, an institution designed to foster knowledge, critical thinking, and civic responsibility.

In his writings, Jefferson expressed a clear vision: an educated populace was essential to the nation's well-being and security. He argued that knowledge was "the common property of mankind" and that education should be universally accessible (Jefferson, 1818). His commitment to education as a democratic right illustrates his belief in its role for national stability and advancement.

Abraham Lincoln
Champion of Education and Innovation

Although Abraham Lincoln lived in a later era, his contributions to the nation's educational landscape were profound. Lincoln recognized that education could bridge social divides and foster a skilled workforce.

During his presidency, he signed the Morrill Act of 1862, which established land-grant colleges across the United States. This act aimed to provide higher education in agricultural and mechanical arts, promoting skills that would drive economic growth and support national security.

Lincoln's legacy as a champion of education laid the groundwork for modern public education in America. His dedication to creating educational opportunities for all aligned with the Founding Fathers' vision of knowledge as a national asset. Through education, Lincoln sought to empower citizens with the tools needed to contribute to society, ensuring that the nation could withstand both internal and external challenges.

Linking the Founding Vision to National Security

The Founding Fathers' collective vision for education was rooted in their understanding that an informed society is better equipped to navigate challenges and safeguard its future. Their support for science, technology, and practical knowledge foreshadowed the essential role that STEMM education now plays in national security.

In a modern context, the fields of science, technology, engineering, mathematics, and medicine are instrumental in defending the nation, from advancements in medical sciences to innovations in cyber defense and military technology.

The lessons from Washington, Franklin, Jefferson, and Lincoln show that an investment in education is an investment in national stability. By fostering scientific inquiry and creating institutions for learning, they set the stage for a nation that values and protects knowledge. Today, their vision is reflected in STEMM initiatives that support the country's defense and position it as a global leader in science and technology.

The Founding Fathers understood the transformative power of education and its critical role in shaping a resilient nation. From Washington's agricultural innovations to Franklin's scientific breakthroughs, these early leaders championed knowledge and education as essential to national prosperity and security. Their vision for an educated, scientifically minded society continues to inspire current efforts in STEMM education, underscoring that a commitment to knowledge and innovation is fundamental to safeguarding the nation's future.

Chapter 4
Beyond the Classroom
&
STEMM in the Real World

STEMM education extends far beyond the classroom. While foundational learning occurs in schools, the true impact of STEMM is often realized through real-world applications, internships, and hands-on experiences that prepare students for a dynamic workforce. In this chapter, we will explore how practical experiences and applied learning bridge the gap between theoretical concepts and real-world challenges, giving students the tools they need to thrive in STEMM fields and contribute to national progress.

The Power of Hands-On Learning

Hands-on learning is a vital part of STEMM education. By engaging in fun scientific experiments, scientific projects, and science activities, students can apply what they've learned to practical scenarios, helping to solidify their understanding. Research consistently shows that students learn more effectively when they can interact with their subjects, rather than only studying them through textbooks (Jones, 2022). In STEMM, this means going beyond theory to practice engineering in robotics clubs, perform scientific experiments in labs, or build medical skills in simulated patient care, and participating in scientific presentations with their peers.

For example, students interested in biology may benefit from working with real DNA samples or learning lab protocols, which gives them a direct understanding of biotechnology techniques.

Similarly, engineering students can gain critical experience by designing prototypes or working with computer-aided design (CAD) software, equipping them with skills that are essential in professional environments. These hands-on experiences not only make learning more engaging but also prepare students for the technical demands they'll face in their careers.

Connecting Students with Industry for Internships
Internships are an invaluable way for students to experience the STEMM fields firsthand and understand the practical applications of their education. Internships allow students to apply their classroom knowledge to solve real-world problems and come up with solutions. Internships often occur within a collaborative team setting making it more engaging for the student. Through these experiences, students can gain insight into various STEMM professions and develop a clearer sense of their career paths. Additionally, internships help students build professional networks, which can lead to mentorship opportunities and job offers down the workforce pipeline.

Programs like the **National Science Foundation's Research Experiences for Undergraduates (REU)** provide students with hands-on research opportunities in various STEMM fields. These internships not only expose students to cutting-edge scientific research but also allow them to collaborate with professionals, ask questions, and develop practical skills.

Many internships, including those in biotechnology companies or engineering firms, involve projects directly relevant to current societal needs, from developing medical devices to innovating renewable energy solutions.

Real-World Projects and Problem-Based Learning
Problem-based learning (PBL) is a teaching method that challenges students to work on real-world problems, often in teams. This approach helps students develop critical thinking and problem-solving skills, as they must analyze situations, think creatively, and collaborate with others to find solutions. Many STEMM programs incorporate PBL to give students experience with real-world challenges, preparing them for the complexities of the professional world. For instance, engineering students might be tasked with designing a sustainable building or developing a device that could improve healthcare accessibility.

In fields like environmental science, students might work on projects that address local ecological issues, such as pollution in rivers or conservation efforts for endangered species. By working on these problems, students learn how to approach issues from multiple perspectives, gaining skills in both scientific analysis and communication.

Programs That Bridge Education and the Workforce

Numerous programs help bridge STEMM education and the workforce, preparing students to tackle real-world challenges upon graduation.

For example, Girls Who Code is an initiative that introduces middle and high school students to computer science, encouraging them to build websites, apps, and even prototypes that solve community problems. This program not only gives students insight into coding and engineering but also fosters teamwork and problem-solving skills. Biotechnology students can benefit from programs like the Amgen Biotech Experience (ABE), which provides high school students with hands-on laboratory experiences, allowing them to conduct real-world biotech experiments such as DNA sequencing and genetic engineering. These experiences introduce students to lab techniques used in the biotechnology industry, helping them build foundational skills in molecular biology and laboratory science. Other programs, like Project Lead The Way (PLTW), offer specialized courses and experiences for K-12 students, focusing on pathways in engineering, biomedical science, and computer science. By participating in these programs, students gain valuable skills that make them more competitive in the workforce and better prepared for STEMM careers.

Another impactful example is the National Institutes of Health (NIH) Summer Internship Program, which offers research opportunities in biomedical sciences. Students who participate in the NIH program work alongside leading scientists, gaining exposure to medical research and learning about the daily work of professionals in the healthcare industry. Programs like these ensure that students are not only learning in the classroom but are also gaining experience that prepares them for STEMM careers with a clear understanding of real-world expectations.

Bringing Programs and Workshops to Life through Your Contributions

The programs and workshops you've developed or participated in serve as powerful examples of how hands-on learning and real-world applications can elevate STEMM education. By organizing STEMM camps, STEMM seminars, or workshops, you provide students with opportunities to build practical skills, fostering their curiosity through experimentation and creativity. Initiatives like free coloring pages for kids or STEMM outreach programs for underrepresented students make STEMM concepts accessible, inspiring students to dive deeper into exploration and discovery. These efforts help bring STEMM education to life, encouraging students to actively engage and connect with these fields in meaningful ways.

Preparing for a Global Workforce

Real-world STEMM experiences not only prepare students for domestic careers but also give them the skills they need to compete globally. As technology advances rapidly and industries become more interconnected, STEMM professionals must be adaptable, able to work with diverse teams, and skilled in solving complex problems. Science internships, projects, and programs that emphasize global challenges, such as climate change or public health crises, prepare students to think beyond borders and to contribute to solutions that benefit communities worldwide.

Scientific programs that connect students with international organizations or provide opportunities to study abroad can be particularly beneficial. By working on global projects or interacting with international professionals, students broaden their perspectives and learn to approach problems with a comprehensive, globally informed mindset. This adaptability is essential as they enter the workforce and tackle issues that extend beyond national boundaries.

STEMM education becomes truly transformative when students take their knowledge beyond the classroom. Through hands-on learning, internships, real-world projects, and innovative programs, students gain the skills needed to succeed in STEMM careers and make meaningful contributions to society.

By participating in programs that expose them to real-world applications, students can translate academic concepts into practical skills, preparing them for a workforce that increasingly values adaptability, problem-solving, and interdisciplinary collaboration. In this way, STEMM education not only builds individual careers but also strengthens communities and advances national progress.

Chapter 5
Partnerships in Education and Innovation

Partnerships between academic educational institutions, government agencies, and the private sector play an essential role in developing a strong STEMM education system in the United States. These collaborations create pathways for students, researchers, and professionals to share knowledge and resources, accelerating advancements in scientific innovation, national defense, and public health. Chapter five, will explore how these partnerships enhance STEMM education, prepare students for real-world applications, come up with solutions for real-world problems, and drive progress in key areas critical to society.

The Role of Educational Institutions in STEMM Partnerships

Educational institutions are at the heart of STEMM partnerships, providing the foundational knowledge and research skills that drive innovation. Universities and colleges often collaborate with industry leaders to offer students hands-on experience through internships, research programs, and mentorship. For example, many universities partner with biotechnology firms to give students hands on lab experience, training them in cutting-edge techniques that are directly applicable to the workforce. These opportunities make education more relevant to the students' future careers. These applied techniques support bridging the gap between theoretical knowledge and practical applied skills.

Additionally, collaborations between K-12 schools and tech companies help introduce younger students to STEMM fields early on keeping them engage in science. Programs like Apple's ConnectED initiative bring technology resources into underserved schools, enhancing access to digital tools and teaching resources. By working with tech companies, schools can provide students with early exposure to STEMM, sparking interest in these fields and encouraging diverse participation.

Government and Industry Partnerships for National Defense

Government agencies, particularly in defense, have recognized the importance of partnering with educational institutions and private companies to develop a workforce pipeline skilled in STEMM. The Department of Defense (DoD), for example, supports STEMM education through initiatives like the SMART (Science, Mathematics, and Research for Transformation) Scholarship-for-Service Program. This program provides funding for students pursuing degrees in critical STEMM fields in exchange for working with the DoD upon graduation. By investing in students' education, the DoD ensures that highly skilled individuals are prepared to contribute to national defense.

In addition to education funding, government and private sector collaborations drive innovation in defense technologies. Through programs like the Defense Advanced Research Projects Agency (DARPA), the government works closely with universities and private research labs to develop advanced technologies for national security.

DARPA-funded projects have led to groundbreaking innovations, from early internet technology to modern AI applications. These partnerships provide students and researchers with opportunities to work on projects that not only benefit the country's security but also push the boundaries of technological advancement.

Public-Private Collaborations in Public Health

The COVID-19 2020 pandemic underscored the need for partnerships between educational institutions, government, and the private sector in the field of public health. During the pandemic, the collaboration between universities, health agencies, and pharmaceutical companies accelerated vaccine development, highlighting the critical role of partnerships in responding to global health crises. The Moderna vaccine, for instance, was developed with contributions from the National Institutes of Health (NIH) and private pharmaceutical companies. By combining research expertise, resources, and distribution networks, these partnerships enabled a rapid response that saved millions of lives. Public health partnerships also extend beyond crisis response, playing a long-term role in health innovation. Universities and medical institutions often work with private companies to conduct research and develop new medical technologies. For example, partnerships between hospitals and tech firms have led to advancements in telemedicine, making healthcare more accessible to rural and underserved populations. By pooling resources and expertise, these partnerships improve healthcare outcomes and expand the reach of medical services.

Innovations from Cross-Sector STEMM Initiatives

Cross-sector STEMM initiatives drive innovation by bringing together diverse perspectives and expertise. Programs like the National Science Foundation's Industry-University Cooperative Research Centers (IUCRC) encourage collaboration between academia and industry to address shared research goals. Through these partnerships, university researchers work with industry representatives to identify real-world problems and develop solutions that can be directly applied in the field. These initiatives not only foster technological advancement but also provide students with valuable experience, making them more competitive in the workforce.

Another example of cross-sector innovation is the National Institute for Innovation in Biomanufacturing and Advanced Bioengineering (NIIBAB), a public-private consortium focused on advancing biotechnologies for healthcare, environmental, and industrial applications. By bringing together experts from government, academia, and industry, NIIBAB addresses critical challenges in bioprocessing, bioengineering, and sustainable biomanufacturing. This collaboration supports the development of novel solutions with applications in areas such as personalized medicine, environmental restoration, and bio-based materials, ensuring that the United States remains at the forefront of biotechnological and bioindustrial advancements in the global arena.

Building a Workforce Ready for Tomorrow's Challenges

These partnerships between education, government, and industry do more than advance science and technology. These partnerships are instrumental in building a workforce that is ready for tomorrow's challenges. When students gain experience through partnerships with companies, research institutions, and government programs, they are better prepared to enter the workforce with practical applied skills and an understanding of real-world applications. Additionally, industry and government benefit from these partnerships by accessing a workforce pipeline of skilled STEMM-educated workers who can contribute to their missions and objectives.

Programs such as Pathways to STEM Excellence and the National Math and Science Initiative (NMSI) work with schools to provide students from diverse backgrounds with STEMM education and mentorship. By investing in these partnerships, the U.S. builds a workforce that reflects the diversity and creativity of its population, a critical factor in fostering innovation and problem-solving.

An additional example of impactful cross-sector collaboration is STEM for Her, a nonprofit organization dedicated to advancing girls and young women in STEM fields. STEM for Her collaborates with schools, industry leaders, and community organizations to provide hands-on learning experiences, mentorship, and career pathways that inspire young women to pursue careers in science, technology, engineering, and mathematics. By fostering a supportive network and creating access to STEM resources, STEM for Her addresses the gender gap in STEM fields, ensuring that more women contribute to and lead in areas like biotechnology, environmental science, and forensic science. This initiative plays a critical role in shaping a diverse STEM workforce that reflects the broader community and strengthens the nation's capacity for innovation. Through programs and partnerships, STEM for Her empowers the next generation of female scientists and engineers, promoting inclusive growth and diversity across sectors essential to global competitiveness.

Partnerships in STEMM education is essential to building a resilient and innovative society. By connecting educational institutions, government, and the private sector, these collaborations accelerate technological advancements, strengthen national security, and improve public health outcomes. As we look toward the future, the need for strategic partnerships will only grow, creating more opportunities for students and professionals to contribute to critical fields and ensuring that the United States remains a leader in science, technology, engineering, mathematics, and medicine. Through sustained collaboration, these partnerships serve as a cornerstone of progress and play an indispensable role in shaping a brighter future.

Chapter 6
Workforce Development for a Modern Society

As technology advances rapidly and global challenges become more complex, the demand for skilled workers in STEMM fields (Science, Technology, Engineering, Mathematics, and Medicine) continues to grow. To stay competitive and bolster national security, the United States must invest in developing a skilled workforce capable of meeting the demands of a modern society. Chapter 6 will examine the initiatives, programs, and strategies designed to prepare the next generation of STEMM professionals, equipping them with the skills and flexibility essential for success in an ever-evolving world.

Building a Workforce for the Future

In an interconnected global landscape where technology continuously transforms industries, the United States depends on a highly skilled STEMM workforce to tackle critical challenges across essential fields like cybersecurity, healthcare, public health, biotechnology, climate science, and bioengineering. The National Science Board has projected that the number of jobs requiring STEMM skills will continue to grow at a faster rate than other occupations, underscoring the importance of preparing students for careers in these fields (National Science Board, 2023). Developing a workforce capable of navigating this complex environment is essential to sustaining economic growth and securing national interests on a global stage.

Strengthening STEMM Education at Every Level

Preparing a skilled STEMM workforce begins with a strong educational foundation at all levels, from K-12 to higher education. K-12 STEMM education introduces students to the basic principles of science, technology, engineering, mathematics, and medicine, inspiring them to pursue these fields and laying the groundwork for more specialized studies. Programs like Project Lead The Way (PLTW) and NASA's STEM Engagement initiatives bring STEMM concepts to life through hands-on learning experiences and problem-solving activities. By fostering curiosity and critical thinking from an early age, these programs help students build a passion for STEMM disciplines and develop the foundational skills they will need in higher education and beyond.

In higher education, colleges and universities offer specialized programs that prepare students for specific STEMM careers. Community Colleges also play a significant role in workforce development by offering flexible and affordable STEMM training programs. Partnerships between community colleges and local industries can create tailored training pathways, equipping students with the technical skills needed in their local job markets.

Programs like the Advanced Technological Education (ATE) initiative by the National Science Foundation (NSF) support these collaborations, allowing students to gain hands-on experience in fields like advanced manufacturing, biotechnology, and environmental technology.

Bridging the Gap Between Education and Industry

One of the greatest challenges in workforce development is bridging the gap between what students learn in classrooms and the skills required in professional settings. To address this, many educational institutions and businesses collaborate to provide students with real-world experience through internships, apprenticeships, and co-op programs. These experiences allow students to apply their knowledge in a professional environment, giving them practical skills and a deeper understanding of industry demands. For example, the Pathways to STEM Excellence program partners with major technology companies to offer students internships and mentorship opportunities. Through this program, students work alongside industry professionals on projects that have real-world applications, such as software development, data analysis, and research. Similarly, apprenticeship programs in fields like advanced manufacturing and cybersecurity give students hands-on training while they are still in school, allowing them to gain valuable work experience and build connections in their chosen fields.

Encouraging Lifelong Learning and Skill Development

Given the rapid pace of technological change, developing a future-ready workforce requires a commitment to lifelong learning. In the past, formal education may have been enough to support a person's entire career. However, in the modern world, workers in STEMM fields must continually update their skills to keep up with new advancements. This need has led to an increase in continuing education programs and professional development opportunities.

Take advantage of free educational resources and organizations like Khan Academy, MIT OpenCourseWare, and Harvard Online Courses, which offer online courses that allow professionals to learn new skills on their own schedules. These platforms partner with leading institutions to provide courses on topics ranging from data science and artificial intelligence to medical research and environmental science. Additionally, platforms like Coursera, edX, and LinkedIn Learning offer courses and certifications, which, while requiring a fee, can be worthwhile investments for advancing your career. By making education accessible and affordable, these programs support the continuous development of a highly skilled workforce, which is essential for maintaining the United States' competitiveness on a global stage.

The Role of Government in Workforce Development

The U.S. government plays an active role in workforce development, particularly in fields critical to national security. Agencies such as the Department of Defense (DoD) and the National Institutes of Health (NIH) invest in programs that support STEMM education and training. For instance, the DoD's SMART Scholarship-for-Service Programprovides funding for students pursuing degrees in critical STEMM fields, such as engineering and cybersecurity, in exchange for service with the DoD after graduation. By investing in education, the government helps ensure that it has a steady pipeline of talent equipped to meet the challenges of the modern world.

Similarly, the National Science Foundation (NSF) supports STEMM workforce development through initiatives like Innovative Technology Experiences for Students and Teachers (ITEST). This program funds projects that introduce middle and high school students to technology-intensive fields, aiming to build a workforce that can address the nation's scientific and technological needs. Government funding and policy support for these programs are essential in strengthening the pipeline of future STEMM professionals.

Building a Broadly Skilled STEMM Workforce

A workforce composed of individuals with varied perspectives, experiences, and problem-solving approaches is crucial for driving innovation. To develop a strong and adaptable STEMM workforce, it's essential to foster opportunities at all levels of education and in the workplace. Programs like the National Math and Science Initiative (NMSI) and Girls Who Code work to remove barriers and create pathways for individuals who have been historically underrepresented in STEMM fields. By offering resources, mentorship, and support, these programs help students from a wide range of backgrounds thrive and succeed in STEMM disciplines. Encouraging diversity in STEMM also includes addressing gender and socioeconomic disparities. Initiatives that provide scholarships, mentorship, and outreach programs for women and underrepresented minorities help create a workforce that reflects the full spectrum of society. This inclusivity fosters a more innovative and adaptive STEMM community, equipped to handle the complex and evolving challenges of a global society.

Preparing for Future Challenges

As we look to the future, the demand for a skilled STEMM workforce will continue to expand. Emerging fields like artificial intelligence, biotechnology, renewable energy, and biomanufacturing will need professionals with specialized expertise and flexible, adaptable skills. Areas such as quantum computing, precision medicine, gene editing, cybersecurity, and space exploration are evolving rapidly and will require new talent prepared to address complex challenges. Developing a workforce capable of advancing these fields is essential to maintaining the United States' leadership in innovation, security, and global influence. Programs that promote applied skills, industry experience, and lifelong learning are vital in this endeavor. By investing in these initiatives, the U.S strengthens its ability to respond to future challenges, from healthcare advancements to environmental sustainability. Preparing the workforce of tomorrow is not only a matter of individual success but also a critical component of the nation's ability to thrive in an interconnected world. Workforce development is fundamental to a modern, secure, and competitive society. Through a combination of early STEMM education, industry partnerships, lifelong learning, government support, and a commitment to diversity, the United States is creating a workforce capable of meeting the demands of the future. As global challenges continue to evolve, so must the strategies for preparing our workforce. This will ensure that STEMM professionals are equipped to drive innovation and uphold national security on a global stage. By investing in workforce development, the nation not only secures its economic stability but also fortifies its position as a leader in the fields that will shape the future.

Chapter 7

The Role of Scientific Societies

Scientific societies are foundational to advancing STEMM fields (Science, Technology, Engineering, Mathematics, and Medicine) through promoting research, setting professional standards, and fostering a collaborative scientific community. Since the early days of American science, societies such as The Washington Academy of Sciences, the American Association for the Advancement of Science (AAAS), and Sigma Xi, The Scientific Research Honor Society, have been instrumental in establishing networks that allow scientists, engineers, medical professionals, and researchers to share knowledge and collaborate on projects that push innovation forward.

Chapter seven explores the importance of these societies in setting ethical and professional standards, encouraging groundbreaking discoveries, and nurturing the next generation of STEMM professionals. Beyond knowledge sharing, these societies provide mentorship and training opportunities that help young scientists navigate their careers in increasingly complex fields.

They also serve as a voice for science in policy discussions, ensuring that scientific insights are integrated into public decision-making. By advocating for public science literacy, these organizations play a key role in strengthening society's trust in scientific advancements. Through their conferences, publications, and outreach programs, these societies continue to drive progress across disciplines. Their enduring influence reflects a commitment to both scientific excellence and societal impact, creating a legacy that advances STEMM education and innovation.

The Washington Academy of Sciences
Leading America's Scientific Legacy

Founded in 1898, The Washington Academy of Sciences was one of the earliest scientific societies in the United States, established to support scientific research and facilitate collaboration among scientists. The Academy's mission was to create a scientific community within the nation's capital, providing a platform for the exchange of knowledge across various scientific disciplines. The Washington Academy of Sciences still plays a crucial role in advancing STEMM fields, serving as a hub for scientists, policymakers, and educators who work collectively to address pressing issues. Through its network of affiliates, the Academy promotes interdisciplinary research and provides educational resources to foster public understanding of science. By hosting annual conferences, publishing scientific journals, and offering professional development opportunities, the Academy strengthens connections within the scientific community. Its legacy of bringing together diverse scientific fields continues to shape STEMM education and influence policy in the nation's capital and beyond. Today, it remains a beacon for scientific progress, fostering a collaborative spirit that is essential for addressing the complex challenges of the modern world. Its commitment to fostering curiosity and critical thinking extends to outreach programs that inspire the next generation of scientists and innovators. The Academy's ongoing influence underscores the importance of science in shaping a secure and innovative future for society.

The American Association for the Advancement of Science (AAAS)
Bridging Science and Society

The American Association for the Advancement of Science (AAAS), founded in 1848, is one of the most influential scientific societies globally, dedicated to advancing science and innovation for the benefit of all people. AAAS established the prestigious journal Science, providing a platform for groundbreaking research across a wide range of disciplines. With a strong commitment to science communication, AAAS strives to bridge the gap between complex scientific concepts and public understanding. It offers programs to train scientists in effective communication and policy engagement, empowering them to convey their work to the public and policymakers alike.

During critical moments in history, AAAS has been a leading voice advocating for scientific research, funding, and policy, ensuring that scientific knowledge informs public debate and decision making. The organization's emphasis on public outreach has been vital in shaping public understanding of science, especially during crises like the COVID-19 2020 pandemic. By advancing scientific literacy and supporting interdisciplinary research, AAAS upholds its mission to promote knowledge and foster collaboration among scientists, educators, and the public.

Sigma Xi Research Honor Society
A Commitment to Research Excellence and Integrity

Founded in 1886, Sigma Xi, The Scientific Research Honor Society, is dedicated to recognizing and supporting excellence in scientific research. As one of the oldest and most respected scientific societies, Sigma Xi promotes high ethical standards and rigorous research practices across various fields. The society's Grants-in-Aid of Research (GIAR) program funds research projects by undergraduate and graduate students, fostering the next generation of scientists. This program ensures that young researchers have the resources they need to pursue their ideas, contributing to innovation in STEMM fields. Sigma Xi's commitment to ethical research practices extends beyond its members. The organization advocates for responsible scientific conduct and transparency across the scientific community.

By providing a platform for scientists to engage in discussions on research integrity, Sigma Xi plays a vital role in upholding the credibility of scientific findings. This emphasis on ethics and excellence is essential in building public trust in science, especially in an era of rapid technological change and information sharing. Through its international reach and partnerships with other scientific organizations, Sigma Xi amplifies its mission of promoting rigorous, ethical research globally. Additionally, the society's conferences and publications offer a space for sharing groundbreaking research and fostering interdisciplinary collaboration.

Fostering Innovation and Advancing Knowledge

In addition to upholding standards, scientific societies create opportunities for collaboration, enabling professionals across disciplines to share ideas, explore new concepts, and push the boundaries of innovation. Organizations like the American Chemical Society (ACS), American Association for the Advancement of Science (AAAS), and the IEEE (Institute of Electrical and Electronics Engineers) host annual conferences, workshops, and publish peer-reviewed journals, advancing knowledge in their respective fields. ACS, one of the world's largest scientific societies, promotes the chemical sciences, while IEEE is central to progress in electrical engineering, computing, and electronics. By facilitating knowledge exchange, these societies play an active role in fostering scientific and technological advancement.

These events provide scientists, engineers, and researchers with platforms to present their work, gain feedback, and collaborate on projects with real-world impact. For instance, IEEE's conferences on emerging technologies, like artificial intelligence and cybersecurity, draw professionals from around the world, fostering innovation in fields that are critical to national security and global competitiveness. These gatherings also provide educational workshops and professional development opportunities, equipping participants with the latest tools and techniques in their fields. Through their extensive networks, these societies help bridge the gap between academic research and industry applications, ensuring that scientific discoveries reach their full potential in addressing societal needs.

Supporting Young Scientists and Career Development

Scientific societies have long recognized the importance of supporting early-career scientists. Many organizations offer scholarships, grants, and mentorship programs to help students and young researchers pursue STEMM careers. For example, Sigma Xi's Grants-in-Aid of Research and AAAS's science policy fellowships provide funding and support for students to conduct research, gain professional experience, and explore different career paths.

Additionally, societies like ACS and IEEE provide career resources, such as job boards, skill-building workshops, and networking events. By connecting early-career scientists with mentors and industry professionals, these organizations help bridge the gap between education and the workforce, empowering young scientists with the tools and guidance they need for successful careers. This support system is critical in building a skilled workforce that can adapt to the demands of modern society and contribute to advances in science and technology.

Furthermore, these opportunities provide young scientists with exposure to interdisciplinary collaboration, equipping them to tackle complex global challenges. By nurturing talent from the ground up, scientific societies play a vital role in sustaining innovation and ensuring the continued growth of STEMM fields across sectors.

Promoting Public Understanding of Science

Public understanding of science is crucial for informed decision-making and societal progress. Organizations like AAAS prioritize science communication and outreach to ensure that scientific information is accessible and accurate. Through initiatives like AAAS's Center for Public Engagement with Science and Technology, scientists receive training in communicating their research to the public, helping to demystify complex topics and promote evidence-based understanding.

Efforts to promote public understanding are especially important during health and environmental crises, where misinformation can have serious consequences. During the COVID-19 pandemic, societies like AAAS and the American Public Health Association (APHA) worked to provide accurate and timely information to the public.

Advocating for sound science and countering misinformation helped guide public behavior, support informed decision-making, and build trust in scientific institutions during a critical time. In addition to addressing misinformation, these organizations also engaged communities by providing resources to understand and manage health risks effectively. Public outreach campaigns, online information hubs, and virtual town halls were launched to foster transparency and keep citizens informed. These efforts underscore the essential role scientific societies play in bridging the gap between researchers and the public, empowering communities to make informed choices in times of crisis.

Building a Global Scientific Community

Scientific societies also play a vital role in building an international network of researchers, facilitating cross-cultural collaboration, and addressing global issues. Organizations such as the American Geophysical Union (AGU) and the International Union of Pure and Applied Chemistry (IUPAC) bring together scientists from around the world to share research and develop solutions to shared challenges. These partnerships are especially important for addressing global concerns like climate change, infectious diseases, and sustainable energy.

The global focus of these societies fosters an inclusive scientific community that values diverse perspectives and approaches, strengthening international cooperation. By supporting global partnerships, scientific societies not only advance research but also build a more connected, resilient scientific community capable of responding to international challenges. Conferences, publications, and joint research initiatives sponsored by these societies provide valuable opportunities for scientists to exchange ideas and build lasting professional relationships. Additionally, collaborative efforts can accelerate the pace of discovery, as researchers pool resources, share data, and develop solutions more efficiently. In a world increasingly defined by interconnected challenges, these international collaborations are essential for fostering scientific progress that transcends borders.

Advocating for Research Funding and Science Policy

Scientific societies are often at the forefront of advocating for research funding and policies that support scientific progress. Through collaborations with government agencies and policymakers, societies like AAAS and The Washington Academy of Sciences highlight the importance of investment in STEMM education, research, and innovation. Their advocacy ensures that scientific research remains a national priority, enabling continued advancements in research and development, scientific innovation, public health, environmental conservation, and national security.

By emphasizing the societal and economic benefits of scientific research, these organizations help secure resources that fuel continued progress in STEMM fields. This advocacy strengthens the foundation of scientific research in the United States and ensures that STEMM advancements contribute to both the well-being of society and the country's global leadership. Additionally, these societies provide a vital voice for the scientific community in shaping legislation, ensuring that science-informed policies address critical issues. Through public campaigns and partnerships, they raise awareness about the long-term impacts of sustained research funding on society's quality of life. By bridging the gap between scientists and policymakers, these societies create a pathway for translating research into meaningful action that benefits both current and future generations.

The Legacy and Future of Scientific Societies

Scientific societies such as The Washington Academy of Sciences, AAAS, and Sigma Xi have built legacies of advancing knowledge, fostering collaboration, and upholding ethical standards. These organizations continue to adapt to the evolving needs of the scientific community, supporting the next generation of innovators and ensuring that science serves society's needs.

Looking forward, these societies will remain essential to the advancement of STEMM fields. As new challenges emerge, they will continue advocating for responsible research, supporting diverse voices, and promoting public understanding of science. Their commitment to advancing science and fostering collaboration ensures that STEMM disciplines will keep evolving to meet the needs of an ever-changing world. Scientific societies are foundational to STEMM progress, providing platforms for collaboration, innovation, and the ethical advancement of science and technology. Through setting standards, supporting young scientists, promoting public understanding, and advocating for research funding, these organizations play an indispensable role in shaping the future of STEMM. By uniting scientists and fostering a spirit of discovery, societies like The Washington Academy of Sciences, AAAS, and Sigma Xi help build a world where science enriches society and prepares it to face the challenges ahead.

Their enduring dedication to scientific integrity and societal progress solidifies their role as both stewards of knowledge and champions of a scientifically literate world.

Chapter 8
Biotechnology
The Frontier of Emerging Technologies

Special Section
The Importance of Understanding Biotechnology

Biotechnology is the application of biological principles to create innovative products and solutions and is advancing rapidly. Biotechnology is reshaping fields such as medicine, agriculture, environmental science, national security, and forensic science. Its impact is far-reaching, with applications in diagnostics, genetic modification, biosecurity, and crime scene investigation. Biotechnology addresses some of society's most complex and urgent challenges.

In this special section, Chapter Eight, we will dive into biotechnology's crucial role in improving medical treatments, advancing forensic science, protecting national security, and fueling economic growth. Understanding these advancements will illuminate how biotechnology continues to transform our world, offering solutions that support health, safety, and sustainability for the future.

Moreover, biotechnology's potential for personalized medicine promises to revolutionize patient care by tailoring treatments to individuals' genetic profiles. In agriculture, biotech innovations like genetically modified crops are enhancing food security by improving crop yields and resilience. This chapter will explore how the continuous evolution of biotechnology opens new avenues for solving global challenges and creating a healthier, more sustainable future.

Medical Advancements Through Biotechnology

One of biotechnology's most impactful applications is in healthcare, where it has revolutionized medical diagnostic tools, treatments, and preventative medicine. Innovations in biopharmaceutical drugs derived from biological sources have led to breakthroughs in treating cancers, autoimmune diseases, and genetic disorders. For instance, monoclonal antibodies are engineered to specifically target proteins involved in certain diseases, offering effective treatments for conditions like rheumatoid arthritis and some cancers (National Cancer Institute, 2022).

Further advancements in gene therapy and CRISPR-Cas9 gene editing allow for targeted interventions at the genetic level, potentially curing genetic diseases at their source. These biotechnological innovations have opened promising avenues for treating conditions that previously lacked effective treatments, transforming patient care by tackling diseases at a molecular level. By focusing on genetic causes, these approaches could ultimately reduce healthcare costs by preventing the progression of severe diseases, underscoring the role of biotechnology as a pillar of public health. In addition to treating genetic disorders, biotechnology has enabled the development of more precise and rapid diagnostic tools, which are critical in detecting diseases in their early stages. Technologies like liquid biopsy and next-generation sequencing allow for non-invasive, comprehensive disease monitoring, improving patient outcomes through early intervention. As research progresses, biotechnology holds the potential to create entirely new treatment modalities, such as personalized cell therapies, which could reshape the future of medicine.

Forensic Science and Biotechnology Transforming Criminal Investigations

Biotechnology has also brought significant advancements to forensic science with innovative methods for identifying perpetrators and solving crimes. DNA profiling is a cornerstone of modern forensics that relies on biotechnological processes to analyze genetic material collected from crime scenes. With techniques such as polymerase chain reaction (PCR), forensic scientists can amplify tiny samples of DNA, allowing them to create a genetic profile that can link suspects to crime scenes or victims with a high degree of accuracy (Butler, 2015). This capability has transformed criminal investigations, reduced wrongful convictions, and enabled the resolution of cold cases. The application of next-generation sequencing (NGS) has further expanded forensic capabilities. NGS allows for more detailed DNA analysis, enabling forensic scientists to identify specific genetic traits, ancestry, and even predict physical characteristics of unidentified individuals. This technology has proven useful in both criminal investigations and humanitarian efforts, such as identifying disaster victims. By bringing precision and depth to forensic analyses, biotechnology strengthens the justice system and enhances public safety, illustrating how scientific advancements directly benefit society (National Institute of Justice, 2021). Portable DNA analysis tools now allow forensic scientists to analyze samples on-site, which expedites case processing and helps investigators make faster decisions. As these technologies evolve, forensic biotechnology will continue to refine and improve the accuracy and efficiency of criminal investigations worldwide.

Biotechnology in Bioterrorism Defense and Biosecurity

Biotechnology is also crucial in national defense, particularly in protecting against bioterrorism and emerging biological threats. The ability to engineer biological agents has created potential security risks, as pathogens could theoretically be weaponized. Biotechnology also provides the tools necessary to counteract such threats. Advances in biosurveillance and rapid diagnostics allow for the early detection of bioterrorism agents, enabling a swift response to prevent widespread outbreaks.

The development of vaccines and antidotes plays a central role in biosecurity. The swift creation of mRNA vaccines during the COVID-19 pandemic highlighted the speed and adaptability of biotechnological solutions. Such rapid responses are crucial in containing threats and protecting public health, as seen with the Department of Defense's partnerships with biotechnological firms to develop countermeasures for potential bioterrorism threats.

Additionally, advancements in portable diagnostic devices enable real-time detection in field settings, providing immediate data to responders in high-risk scenarios. Research into broad-spectrum antiviral agents is also expanding, offering solutions that could quickly neutralize a range of biological agents, further fortifying biosecurity defenses (Department of Defense, 2023).

Environmental Biotechnology and National Security

Biotechnology's applications extend beyond health and safety, addressing environmental concerns that also impact national security. Bioremediation uses microorganisms to clean up pollutants from contaminated sites, such as soil and water affected by oil spills, heavy metals, and other toxins. By employing microbes to degrade harmful substances, bioremediation helps protect ecosystems and communities, mitigating the health risks associated with pollution and contributing to a stable environment (Environmental Protection Agency, 2023).

Agricultural biotechnology also supports national security by ensuring food security through the genetic modification of crops. These genetically modified organisms (GMOs) are engineered for enhanced resilience, increased yields, and reduced reliance on chemical pesticides, which are crucial for maintaining a stable food supply in the face of climate change and global population growth. By securing food resources, agricultural biotechnology reduces the risk of food shortages, thereby contributing to national stability and economic resilience. Furthermore, biotechnology's role in developing drought-resistant and climate-resilient crops is vital for adapting to changing environmental conditions, safeguarding agricultural productivity. Biotechnology also aids in soil conservation efforts by developing crops with root systems that reduce erosion and improve soil health. This comprehensive approach not only enhances food security but also promotes environmental sustainability, benefiting both local and global ecosystems.

Biotechnology as an Economic Catalyst

Biotechnology is not only a scientific frontier but also a major driver of economic growth. The biotechnology sector has expanded rapidly, creating high-skilled jobs and fostering innovation across industries. Biotechnology hubs like California's Silicon Valley, Massachusetts' Biotech Corridor, and North Carolina's Research Triangle have attracted global talent and investment, supported local economies, and contributed to national economic growth.

These clusters are sustained by partnerships between universities, government agencies, and the private sector, which form a network that supports both scientific and economic innovation. Biotechnology startups often receive support from venture capital and government funding, allowing them to transform scientific research into marketable products. For example, CAR-T cell therapy, initially developed in academic laboratories, was commercialized through biotech firms, creating a new avenue for cancer treatment and generating significant economic value (National Science Foundation, 2023). Additionally, advancements in agricultural biotechnology are improving crop resilience, increasing yields, and creating export opportunities, boosting the agricultural sector. Biotech innovation also enhances productivity in industries like pharmaceuticals and biofuels, contributing to a diversified and sustainable economy. As biotechnology continues to grow, it not only strengthens national economic resilience but also solidifies the U.S. position as a leader in global scientific and technological innovation.

Ethical and Regulatory Considerations in Biotechnology

As biotechnology continues to evolve, ethical considerations and regulatory frameworks become crucial for responsible innovation. Technologies like gene editing raise important ethical questions regarding the potential for "designer" genes and the unintended consequences of genetic modifications. Regulatory agencies, including the Food and Drug Administration (FDA) and Environmental Protection Agency (EPA), are responsible for setting guidelines that ensure safety, efficacy, and ethical conduct in biotechnological applications.

The National Institutes of Health (NIH) also addresses these considerations through bioethics programs, which encourage scientists to consider the societal impacts of their work, from privacy concerns in genetic testing to the ecological risks of GMOs. By promoting responsible practices, these agencies and programs balance the benefits of biotechnological advancements with public safety, ensuring that innovations in this field are ethically sound and socially responsible (National Institutes of Health, 2023). Furthermore, international organizations, like the World Health Organization (WHO), work alongside national bodies to establish global standards that guide the safe use of biotechnology. Public engagement is also essential, as informed discussions with communities about the ethical implications of new technologies foster transparency and trust. As biotechnology progresses, continuous evaluation and adaptation of these ethical and regulatory frameworks will be necessary to address new challenges responsibly and sustainably.

The Future of Biotechnology in National Security

Biotechnology will continue to play a transformative role in national security and innovation as threats evolve and technologies advance. Emerging areas such as synthetic biology, which involves designing and constructing new biological parts and systems, offer immense potential for breakthroughs in medicine, sustainable materials, and environmental solutions. However, synthetic biology also presents new security risks, underscoring the importance of developing comprehensive regulatory frameworks to prevent misuse. Investing in biotechnology education and workforce development is essential for maintaining the United States' competitive edge. By training the next generation of scientists in responsible and innovative biotechnological practices, the nation can ensure that this field continues to advance ethically and effectively. Partnerships between government, academia, and industry will be crucial in creating an environment where biotechnology can flourish, meeting societal needs while navigating the ethical and security challenges that accompany rapid technological progress. Biotechnology stands at the frontier of emerging technologies, offering solutions that impact healthcare, environmental protection, national security, and forensic science. Its contributions to medical advancements, biosecurity, environmental sustainability, and justice illustrate biotechnology's vital role in modern society.

Chapter 9
The Future of STEMM and Innovation

As society advances into the 21st century, emerging technologies are reshaping industries, challenging traditional methods, and opening new frontiers in STEMM (Science, Technology, Engineering, Mathematics, and Medicine). Innovations in fields like artificial intelligence, quantum computing, and bioengineering are transforming the way we live and work, but they also present complex challenges that require adaptable education systems and workforce strategies. Chapter nine explores these emerging trends and highlights the crucial role of STEMM education in preparing students for a rapidly evolving landscape, where technological innovation is increasingly linked to national security, economic resilience, and social progress.

The growing interdependence between STEMM fields also demands interdisciplinary education approaches, where students learn to apply knowledge across various scientific and technological domains. This chapter examines how collaborative learning environments and hands-on experiences can equip students with critical thinking and problem-solving skills essential for tackling modern challenges. Additionally, as these advancements reshape workforce needs, STEMM education must also emphasize adaptability, preparing students not only for current technologies but also for those yet to be imagined.

Artificial Intelligence
The Next Frontier in Automation and Decision-Making

Artificial Intelligence (AI) is one of the most impactful technologies on the horizon, with applications spanning healthcare, cybersecurity, environmental science, and even space exploration. Through machine learning algorithms, AI systems can process vast amounts of data, recognize patterns, and make informed decisions with minimal human intervention. AI's potential to improve decision-making is especially valuable in national security, where it can be used for cybersecurity, threat detection, and intelligence analysis (Goodman, 2022).

In healthcare, AI-powered diagnostic tools assist doctors in identifying diseases with greater speed and accuracy. For example, AI algorithms can analyze medical images for early signs of conditions like cancer or Alzheimer's disease, allowing for earlier intervention and improving patient outcomes (National Institutes of Health, 2023). Similarly, AI is revolutionizing bioinformatics, where it helps process genomic data to identify genetic markers of diseases, supporting the development of personalized medicine.

However, AI also presents ethical and security challenges. As AI becomes more integrated into critical systems, concerns about data privacy, bias, and autonomous decision-making grow. Malicious actors could exploit AI systems, highlighting the need for robust cybersecurity measures. This complexity requires STEMM education to focus not only on developing technical skills but also on ethics, critical thinking, and an understanding of AI's societal implications.

Quantum Computing
Transforming Data Processing and Security

Quantum computing is a field that leverages quantum mechanics to perform computations at speeds far beyond current classical computers, has transformative potential in areas like cryptography, materials science, and artificial intelligence. Unlike traditional computers that process data in binary, quantum computers use quantum bits (qubits), which can exist in multiple states simultaneously. This allows quantum computers to solve complex problems that would take classical computers thousands of years to compute (National Science Foundation, 2023).

One of the most promising applications of quantum computing is in cryptography. Quantum computers could break many of today's encryption methods, posing both a risk and an opportunity for national security. While this capability could potentially render current data encryption obsolete, it also opens the door to developing quantum encryption methods that are far more secure than existing standards. Governments worldwide, including the United States, are investing in quantum research to stay competitive and secure, leading to a growing demand for professionals with specialized knowledge in quantum mechanics, mathematics, and computer science. In medicine, quantum computing could accelerate drug discovery by simulating molecular interactions at a scale previously unimaginable. This capability could lead to faster development of treatments for complex diseases, showcasing quantum computing's potential to revolutionize healthcare.

Bioengineering and the Rise of Synthetic Biology

Bioengineering, including synthetic biology, is a rapidly advancing field that combines biology and engineering to design and construct new biological systems. This innovation has applications in areas ranging from healthcare to environmental conservation and biomanufacturing. Synthetic biology allows scientists to engineer organisms for specific purposes, such as producing biofuels, breaking down environmental pollutants, or manufacturing pharmaceuticals at scale (National Academies of Sciences, Engineering, and Medicine, 2022).

One of the most revolutionary applications of synthetic biology is in regenerative medicine, where bioengineers are working on growing human tissues and organs in laboratories. This research could address organ shortages and transform transplantation procedures, giving hope to patients who need life-saving transplants.

The implications of synthetic biology for national security are also significant. Engineered organisms could theoretically be used for bioterrorism, making biosecurity a top priority as these technologies advance. Education systems must therefore include biosecurity training and ethics as part of bioengineering curricula, ensuring that future scientists are equipped to innovate responsibly and understand the potential risks associated with synthetic biology.

Robotics and Automation Redefining the Workforce

The integration of robotics and automation into various industries is transforming the workforce, creating new opportunities and challenges. Robotics has applications across manufacturing, healthcare, logistics, and even space exploration, where robots are used for tasks that are too dangerous or complex for humans. In healthcare, robotic-assisted surgeries allow for minimally invasive procedures, improving patient recovery times and outcomes (American Medical Association, 2023).

Automation is also reshaping fields like agriculture, where autonomous vehicles and drones monitor crops, manage irrigation, and improve yields. As robots and automation become more common, the demand for skilled workers who can design, program, and maintain these systems is increasing. Education systems must adapt to include robotics and automation in STEMM curricula, preparing students for careers in these expanding fields.

However, the rise of automation brings concerns about job displacement. Preparing students for this future means equipping them with adaptable skills, including programming, systems engineering, and problem-solving, to ensure they can thrive in an automated economy. Robotics courses, internships, and hands-on learning experiences can help bridge the gap between classroom knowledge and real-world applications, fostering a workforce ready for tomorrow's challenges.

Cybersecurity
Defending Digital Infrastructure

As reliance on digital systems grows, so does the importance of cybersecurity. Protecting information systems from cyber threats is vital to maintaining national security, economic stability, and public trust. From personal data breaches to state-sponsored cyberattacks, the demand for skilled cybersecurity professionals has never been higher. Cybersecurity training programs, often supported by partnerships between educational institutions, tech companies, and government agencies, are essential for developing a workforce equipped to defend against evolving threats (Department of Homeland Security, 2023).

Cybersecurity encompasses network defense, ethical hacking, and secure software development. In an era where AI and quantum computing can be used to exploit vulnerabilities, cybersecurity training must be continuously updated to reflect the latest advancements. STEMM education systems must prioritize cybersecurity skills, offering courses in digital forensics, cryptography, and risk assessment to prepare students for roles in protecting the nation's digital infrastructure.

The Role of Education in Preparing for the Future of STEMM

Preparing students for the future of STEMM requires more than traditional instruction; it necessitates an adaptable, interdisciplinary approach that integrates emerging technologies and addresses ethical, environmental, and security implications. Education systems must evolve to provide students with hands-on experiences, project-based learning, and exposure to real-world applications that bridge theory and practice.

Curriculum updates are essential to keep pace with technological advancements. Schools and universities should incorporate AI, quantum computing, and bioengineering into STEMM curricula, offering specialized courses and labs that allow students to engage directly with these fields. Industry partnerships, internships, and mentorship programs can also connect students with professionals, giving them a first-hand look at careers in emerging STEMM fields.

Furthermore, promoting lifelong learning is crucial in each of these fields that evolve as quickly as STEMM. Professional development programs, online courses, and certification pathways allow workers to continually update their skills and adapt to new technologies. By fostering a culture of lifelong learning, education systems can ensure that today's students remain innovative, adaptable, and ready to meet the challenges of tomorrow.

The future of STEMM and innovation lies in embracing emerging technologies and equipping students with the skills to navigate complex, evolving fields. From AI to quantum computing and bioengineering, these advancements hold immense promise for addressing societal challenges, bolstering national security, and driving economic progress. As STEMM fields continue to evolve, so too must education systems adapt, ensuring that students are prepared not only to thrive in these areas but also to innovate responsibly and consider the ethical and security implications of their work. By investing in education and fostering a workforce skilled in emerging technologies, the United States can secure its leadership on the global stage, shaping a future defined by scientific progress, resilience, and responsible innovation.

Chapter 10
Summary

As we have explored in this book, STEMM (Science, Technology, Engineering, Mathematics, and Medicine) is far more than a collection of academic subjects; it is the foundation upon which we build a resilient, prosperous, and secure nation. The importance of investing in STEMM education cannot be overstated. STEMM education equips the next generation with the skills, knowledge, and adaptability needed to tackle the most pressing challenges of our time, from national security to economic growth and public health.

As we conclude this journey, continued commitment to STEMM education and innovation is essential for shaping a future that meets the needs of society and upholds the United States' leadership on the global stage. By fostering curiosity and critical thinking in young learners, we lay the groundwork for breakthroughs that will drive future progress and solutions to global challenges.

This final chapter underscores the responsibility of educators, policymakers, and industry leaders to create opportunities that inspire and empower the next generation. Through sustained investment in STEMM, we not only strengthen our national capacity but also ensure a world where science, technology, and innovation continue to benefit all of humanity.

STEMM Education as a Pillar of National Strength

Investing in STEMM education is, ultimately, an investment in national strength. The ability to innovate, solve complex problems, and adapt to new challenges is critical in today's interconnected world. STEMM trained professionals are at the forefront of fields that safeguard our national security, drive economic prosperity, and improve quality of life. By fostering a generation that is skilled in artificial intelligence, biotechnology, cybersecurity, quantum computing, and other emerging fields, we are positioning the United States to remain competitive and resilient in a rapidly evolving global landscape.

STEMM education contributes not only to individual success but to society's collective strength. A workforce trained in critical STEMM skills is essential for maintaining secure infrastructure, protecting against cybersecurity threats, developing medical advancements, and driving technological breakthroughs. A robust STEMM workforce is integral to national prosperity and safety, creating a society where science and technology address societal needs and empower every community.

The Power of Innovation and Partnerships

Innovation is the driving force that keeps STEMM fields dynamic and responsive to new challenges. As seen through advancements in fields like robotics, AI, and bioengineering, technological progress creates new opportunities and solves real-world problems. However, innovation does not happen in isolation. Partnerships among educational institutions, government, and the private sector are essential for creating the resources, networks, and opportunities that allow students, researchers, and professionals to collaborate effectively and drive STEMM forward.

Programs like the Department of Defense's STEM initiatives, partnerships between universities and tech companies, and public-private collaborations in biotech showcase the transformative power of working together toward shared goals and solutions. Through these partnerships, we can ensure that our education system is responsive to industry needs, that cutting-edge research receives support, and that students gain hands-on experience that prepares them for meaningful STEMM careers. Moving forward, nurturing and expanding these partnerships will be key to ensuring the continued growth of STEMM and to fostering an environment where knowledge and innovation thrive.

Actionable Steps for the Future of STEMM

To build a resilient and innovative future, it is essential to take actionable steps that strengthen STEMM education at all levels. Here are several strategies that can guide future investments in STEMM:

- Early Exposure and Access to STEMM: Introduce STEMM concepts to students early on through K-12 education, focusing on programs that engage underrepresented groups. Programs that provide hands-on learning experiences, such as robotics clubs, science fairs, and coding camps, spark interest and inspire students to pursue STEMM fields.

- Expand Real-World Learning Opportunities: Integrate internships, co-op programs, and apprenticeships into educational curricula to provide students with practical experience. Collaborations between educational institutions and industries can help bridge the gap between classroom learning and workforce needs, preparing students for real-world challenges.

- Promote Lifelong Learning: The rapid pace of technological change demands that STEMM professionals continually update their skills. Encourage a culture of lifelong learning through continuing education programs, online courses, and certifications, making it easier for individuals to adapt to new fields and technologies throughout their careers.

- Invest in STEMM Educator Training: Teachers and Leaders play a critical role in shaping the STEMM pipeline. Investing in STEMM focused educator training programs ensures that teachers and leaders are equipped with the latest knowledge, teaching methods, and tools to inspire students. Support for ongoing professional development is essential to fostering an engaging, inclusive STEMM education environment.

- Support a Broadly Skilled and Representative STEMM Workforce: A workforce with varied perspectives and experiences fuels innovation by enhancing problem-solving approaches. Investing in scholarships, mentorship programs, and outreach initiatives that provide opportunities for women, underrepresented groups, and economically disadvantaged students in STEMM builds a more comprehensive and skilled workforce, ultimately strengthening the STEMM community.

- Encourage Ethical and Responsible Innovation: As technology advances, it is crucial to consider the ethical implications of scientific progress. Education systems should include coursework on bioethics, data privacy, and responsible AI, preparing students to make informed decisions and to use their skills for the benefit of society.

The Path Forward
A Commitment to Progress

To secure the future of STEMM and the nation, it is essential to view STEMM education as a continuous journey rather than a destination. By fostering curiosity, critical thinking, and resilience, we prepare future generations to adapt to new challenges, seize opportunities, and build a society that reflects the best of human ingenuity. Our investment in STEMM education, research, and infrastructure must be ongoing, supported by policies that prioritize innovation, collaboration, and responsible practices. As we move forward, let us remain committed to supporting the partnerships, policies, and educational initiatives that make STEMM accessible to all, ensuring that the promise of science and technology benefits every individual and community. The future belongs to those who are prepared for it, and through dedicated support for STEMM, we can build a future that is as innovative, secure, and resilient as our aspirations demand.

The chapters of this book illustrate the profound impact that STEMM fields have on society and underscore the importance of preparing a workforce ready to lead in a world defined by rapid change and technological advancements. Through STEMM education, we can equip the next generation with the skills and knowledge to drive progress, solve complex problems, and uphold the values that define the United States. By investing in STEMM and fostering a culture of lifelong learning, innovation, and ethical responsibility, we can secure a future where science and technology serve as pillars of national strength, resilience, and prosperity.

In a world where the challenges of tomorrow are being shaped by the technologies of today, our commitment to STEMM will be the guiding light that illuminates a path toward a better future. This journey is a collective endeavor, one that calls on educators, policymakers, scientists, and students alike to envision a world where innovation uplifts us all. Together, through a shared dedication to STEMM education and progress, we can meet the challenges ahead and build a brighter, more secure future for generations to come.

Glossary

Biotechnology: The use of biological processes, organisms, or systems to develop products and technologies that improve human life and the health of the planet.

Education - The process of acquiring knowledge, skills, values, and attitudes through structured learning experiences such as formal schooling, training programs, or self-directed study. In the context of STEMM (Science, Technology, Engineering, Mathematics, and Medicine), education involves not only the theoretical understanding of concepts but also the practical application of these disciplines to solve real-world problems, contribute to innovation, and support societal advancement, particularly in areas like national security, workforce development, and technological progress.

National Security - The protection and preservation of a nation's people, territory, and institutions from threats such as terrorism, espionage, warfare, cyberattacks, and natural disasters. National security involves coordinated efforts across government, military, intelligence agencies, and private sectors to ensure stability, defense, and economic resilience. In the context of STEMM education, national security is increasingly reliant on technological innovation, advanced scientific research, and a well-educated workforce capable of addressing emerging global challenges and safeguarding a nation's interests.

STEMM - Science, Technology, Engineering, Mathematics, and Medicine.

References

Chapter 1

Ferguson, N. M., Laydon, D., Nedjati-Gilani, G., Imai, N., Ainslie, K., Baguelin, M., Bhatia, S., Boonyasiri, A., Cucunubá, Z., Cuomo-Dannenburg, G., Dighe, A., Fu, H., Gaythorpe, K., Green, W., Hamlet, A., Hinsley, W., Okell, L. C., van Elsland, S., Thompson, H., Verity, R., ... & Ghani, A. C. (2020). Impact of non-pharmaceutical interventions (NPIs) to reduce COVID-19 mortality and healthcare demand. Imperial College London. https://doi.org/10.25561/77482

Jackson, L. A., Anderson, E. J., Rouphael, N. G., Roberts, P. C., Makhene, M., Coler, R. N., McCullough, M. P., Crandon, S., Neuzil, K. M., Cohn, A., Clark, J., Makowski, M., Edmonds, K., Follmann, D., Albert, J., Reedy, S., Leav, B., Loddenkemper, J., Meltzer, E., ... & Beigel, J. H. (2021). An mRNA vaccine against SARS-CoV-2—Preliminary report. The New England Journal of Medicine, 383(20), 1920-1931. https://doi.org/10.1056/NEJMoa2022483

Nguyen, L. H., Drew, D. A., Graham, M. S., Joshi, A. D., Guo, C. G., Ma, W., Mehta, R. S., Warner, E. T., Sikavi, D. R., Lo, C. H., Kwon, S., Song, M., Mucci, L. A., Stampfer, M. J., Willett, W. C., Eliassen, A. H., Hart, J. E., Chavarro, J. E., Rich-Edwards, J. W., ... & Chan, A. T. (2020). Risk of COVID-19 among frontline healthcare workers and the general community: A prospective cohort study. The Lancet Public Health, 5(9), e475-e483. https://doi.org/10.1016/S2468-2667(20)30164-X

Chapter 2

Department of Homeland Security. (2023). National Cybersecurity Strategy. DHS Publications.

Goodman, M. (2022). Artificial intelligence and the future of national defense. Journal of Defense Research, 48(3), 235-248. https://doi.org/10.1016/j.defres.2022.07.005

Jackson, L. A., Anderson, E. J., Rouphael, N. G., Roberts, P. C., Makhene, M., Coler, R. N., McCullough, M. P., Crandon, S., Neuzil, K. M., Cohn, A., Clark, J., Makowski, M., Edmonds, K., Follmann, D., Albert, J., Reedy, S., Leav, B., Loddenkemper, J., Meltzer, E., ... & Beigel, J. H. (2021). An mRNA vaccine against SARS-CoV-2—Preliminary report. The New England Journal of Medicine, 383(20), 1920-1931. https://doi.org/10.1056/NEJMoa2022483

National Science Board. (2023). Science & engineering indicators. National Science Foundation.

Smith, J. (2023). Developing tomorrow's workforce: The role of STEM education in national security. Policy and Education Journal, 6(4), 312-324. https://doi.org/10.1080/pej.2023.312

Chapter 3

Franklin, B. (1749). Proposals Relating to the Education of Youth in Pensilvania. University of Pennsylvania Archives.

Jefferson, T. (1818). Report of the Commissioners for the University of Virginia. University of Virginia.

Smith, J. (2023). Founding Visions: How Early American Leaders Shaped Modern Education. Historical Perspectives Press.

Washington, G. (1796). Farewell Address.

Chapter 4

Jones, T. (2022). Hands-on learning in STEMM: Preparing the next generation of innovators. Journal of Educational Research, 45(2), 165-182.

National Institutes of Health. (2023). Summer Internship Program in Biomedical Research. Retrieved from Franklin, B. (1749). Proposals Relating to the Education of Youth in Pensilvania. University of Pennsylvania Archives.

Smith, J. (2023). Connecting STEMM education to the workforce: The power of real-world applications. Education Policy Journal, 12(3), 234-245.

Chapter 5

Department of Defense. (2023). SMART Scholarship-for-Service Program Overview. Retrieved from https://www.smartscholarship.org

National Institutes of Health. (2020). COVID-19 Vaccine Research. Retrieved from https://www.nih.gov

National Institute for Innovation in Manufacturing Biopharmaceuticals. (n.d.). About NIIMBL. Retrieved from https://niimbl.org

National Science Foundation. (2023). Industry-University Cooperative Research Centers Program. Retrieved from https://www.nsf.gov

Chapter 6

National Science Board. (2023). STEM Workforce Data Project. National Science Foundation.

National Institutes of Health. (2023). SMART Scholarship-for-Service Program Overview. Department of Defense.

Project Lead The Way. (2023). STEMM Education Initiatives.

Chapter 7

American Association for the Advancement of Science. (2023). Science Communication Programs. AAAS Publications.

Sigma Xi. (2023). Grants-in-Aid of Research Program Overview. Sigma Xi Publications.

National Science Foundation. (2023). Role of Scientific Societies in STEMM Advancement.

The Washington Academy of Sciences. (2023). History and Mission Overview.

Chapter 8

Butler, J. M. (2015). Advanced Topics in Forensic DNA Typing: Methodology. Academic Press.

Department of Defense. (2023). Biodefense Research and Development Overview. Retrieved from https://www.defense.gov

Environmental Protection Agency. (2023). Bioremediation and Environmental Biotechnology. Retrieved from https://www.epa.gov

Food and Drug Administration. (2023). Biotechnology and Regulatory Science. Retrieved from https://www.fda.gov

National Cancer Institute. (2022). Monoclonal Antibodies in Cancer Therapy. Retrieved from https://www.cancer.gov

National Institute of Justice. (2021). The Future of Forensic DNA Technology. Retrieved from https://nij.ojp.gov

Chapter 9

American Medical Association. (2023). The Role of Robotics in Healthcare. Retrieved from https://www.ama.org

Department of Homeland Security. (2023). Cybersecurity Workforce Development. Retrieved from https://www.dhs.gov

Goodman, M. (2022). Artificial intelligence and the future of national defense. Journal of Defense Research, 48(3), 235-248. https://doi.org/10.1016/j.defres.2022.07.005

National Academies of Sciences, Engineering, and Medicine. (2022). Synthetic Biology: Engineering Life for Societal Impact. National Academies Press.

National Institutes of Health. (2023). Artificial Intelligence in Medical Imaging. Retrieved from https://www.nih.gov

National Science Foundation. (2023). Quantum Computing and the Future of Information Science. National Science Foundation.